*Dedicated to future generations*

*who may be inspired to protect*

*polar bears and their habitat.*

*Blizzard and Glacier's Summer Adventure*
*Story ©2003 Point Defiance Zoo & Aquarium*
*Illustrations ©2003 Gustav Moore*

*Point Defiance Zoo & Aquarium*
*5400 North Pearl Street*
*Tacoma, Washington*
*www.pdza.org*

*Printed in the United States of America by Bang Printing*

*10  9  8  7  6  5  4  3  2  1*

*ISBN: 0-9746321-0-4*

*Join Blizzard and Glacier on an exciting journey to the tundra! You won't want to miss this incredible adventure and all the amazing animals these young bears meet along the way.*

*Visit Point Defiance Zoo & Aquarium on-line at www.pdza.org!*

# Blizzard and Glacier's summer adventure

written by **Robin Bailey**

illustrations by **Gustav Moore**

illustrations by **Robin Bailey and Dan Belting**

story created by **Robin Bailey**

During the early Arctic spring, when it's too cold for you and me to go out to play, two bears wrestled on the Alaskan sea ice.

**"Blizzard and Glacier, stop that now!"** Mom growled.

The young bears were making too much noise. They were scaring away seals that lay nearby. The bears had lived with Mom two whole years... just the right amount of time for baby polar bears and moms to be together.

**Soon, things would change.**

# "Hey, cut it out!

## It's my turn to be next to mom!"

3

4

It was time for Blizzard and Glacier to say goodbye to their Mom and head out on their own adventure! Mom had protected, fed, and taught the twins everything they needed to know. She knew they were ready to hunt and protect themselves.

The cubs asked, **"Are you sure we're big enough?"**

Mom replied, "Yes, it's time. Remember all I've taught you, trust your instincts, and you'll be just fine!"

All animals in nature are born with instincts. Blizzard and Glacier's instincts told them they had to head for land because the sea ice would soon be breaking up in the summer sun.

If they were caught out on the sea ice during the summer, they wouldn't make it to the tundra.

7

As they began their journey toward land, Blizzard spotted a seal pup and its mother lounging by a hole in the ice.

8

"Glacier! Glacier! Look over there, it's a seal pup! This would be a great time to get a bite to eat."

"That seal way over there?" replied Glacier. "Oh, that's too far away for us," but Blizzard lunged off the ice into the water, heading toward the seals.

The seals easily spotted the bears and calmly slipped into the water with a snort and a grunt.

"Smooth move!" yelled Glacier, knowing Blizzard could never catch up with the seals.

Glacier followed Blizzard into the freezing water. The seals were quickly forgotten as the twins began a game of underwater tag. The water didn't feel cold to the bears, because the blubber under their fur kept them warm.

Suddenly,
out of nowhere,
something
barreled by them.

whhooSH!

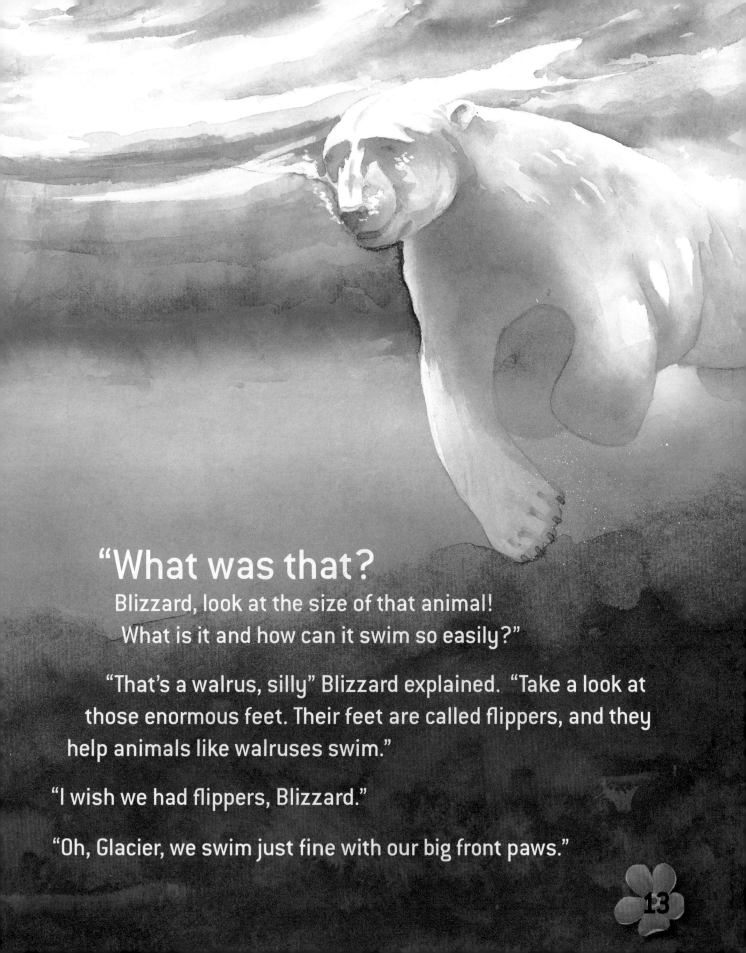

"What was that?
Blizzard, look at the size of that animal!
What is it and how can it swim so easily?"

"That's a walrus, silly" Blizzard explained. "Take a look at those enormous feet. Their feet are called flippers, and they help animals like walruses swim."

"I wish we had flippers, Blizzard."

"Oh, Glacier, we swim just fine with our big front paws."

13

After many more days of traveling over the ice, the bears finally made it to land. It was the beginning of summer, and everything looked different. The ice and snow were disappearing. They could see flowers and lush green grass peeking through the melting snow. **This was the Tundra.**

The bears knew that as a tundra spring turns to summer, it explodes with life. But tundra summers are short, and animals have much to do. Thousands of geese and swans return to nest, and have families. Gigantic herds of caribou forage, or eat, on the new grasses. Arctic hares get fat on tender leaves. Billions of insects hatch, or are born, to fill the sky.

As the bears explored the tundra, Glacier noticed sea otters stuffing their faces with delicious shellfish. The bears chuckled as the otter used its tummy as a dinner table.

**"Look at how much they can eat!"** said Glacier.

**"Otters have to eat many meals to stay warm because they don't have blubber like us,"** Blizzard replied.

**"I'm getting hungry just watching them,"** Glacier said.

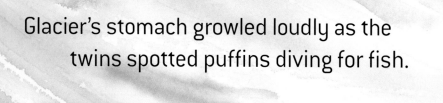

Glacier's stomach growled loudly as the
twins spotted puffins diving for fish.

**"Look at them swim. It's like they're
flying underwater!"** Blizzard said.

"They don't have flippers or strong arms
like us. **They use wings!**" Glacier observed.

Sure enough, the puffins were flying underwater
catching fish with their powerful beaks.
Blizzard's stomach rumbled. **"Now I'm hungry, too!**
Watching these puffins gives me an idea."

19

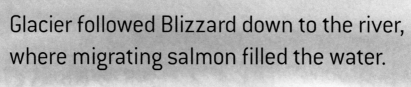

Glacier followed Blizzard down to the river, where migrating salmon filled the water.

"Easy pickings... let's go!"

So they ate...

...and they ate

...and they ate

...until they couldn't eat anymore.

With full tummies, the bears continued exploring the tundra. Just when they thought they'd met every animal possible...

**"Look, a muskox!"** said Blizzard. **"Wow! Check out all that fur. Are they getting shaggier?"**

Sure enough, the muskoxen were growing their long, thick fur coats for the cold tundra winter.

**"It must be time for us to head back to the sea ice,"** Glacier said.

As they neared the water, Blizzard stopped in his tracks. He lifted his nose straight up in the air and his ears perked to attention.

"I smell something, but I don't see it."

They listened and waited, but they didn't have to wait long. It sounded like a deep, loud **shhhh**... The sound hung in the air. Then, they saw who was making the sound in the newly forming sea ice. The beluga whales had finished calving for the summer and were heading for open water with their babies.

As the bears looked back, the whole tundra was changing.

Birds were flying south...

Arctic foxes were growing thick, white, winter fur and puffy tails.

Snowflakes began to fall.

27

The ice was getting thick again. The bears' instincts led them to the edge of the sea.

Winter is a polar bear's time. **Time for adventures... time for feasts on the ice... time for grown-up polar bears to start a family.**

As the snow deepened, Blizzard and Glacier padded off across the ice. They knew this was their time to begin a new adventure.

29

# DiD You KnoW...

**Puffins** use their wings to fly underwater where they catch fish. They always swallow their fish headfirst, tail last.

Female **harp seals** give birth, typically to a single pup born between mid-February to late March. At birth, the pups weigh about 22 pounds.

**Walruses** have 400-700 whiskers on their face that they use to find food. Both male and female walruses have tusks used for hauling out on sea ice and establishing social dominance.

**Muskox** fur is called qiviut (kiv-ee-ute) and is eight times warmer than sheep's wool.

After the snow melts, the **Arctic poppy** grows quickly. The flowers rotate with the sun during the day. The petals become very warm, attracting many insects.

Instead of a dorsal fin the **beluga** has a dorsal ridge running down the center of the back. This ridge helps the whales break through sea ice to create a breathing hole.

**Polar bears** are the largest land carnivores. They are also excellent swimmers and often swim many hours at one time.

**Arctic hares** live in groups of 100-300 animals. Hopping on their hind legs like a kangaroo, they can reach speeds of 64 km per hour.

**Sea otters** are one of the few non-primate animals to use tools, such as rocks to break open shellfish.

**Arctic fox** fur changes color with the seasons. In the wintertime, the fur on their puffy tail, the soles of their feet, and inside their ears helps keep them warm.

Visit **www.pdza.org** for lessons, ideas and further natural history information.